A Little Princess

A Guide for
Teachers and Students

Classics for Young Readers

Sir Gibbie by George Macdonald
Sir Gibbie: A Guide for Teachers and Students
Hans Brinker by Mary Mapes Dodge
Hans Brinker: A Guide for Teachers and Students
A Little Princess by Frances Hodgson Burnett
A Little Princess: A Guide for Teachers and Students

A Little Princess

A Guide for
Teachers and Students

Ranelda Mack Hunsicker

P&R
PUBLISHING
P.O. BOX 817 • PHILLIPSBURG • NEW JERSEY 08865-0817

©2002 by Ranelda Mack Hunsicker

All rights reserved. Except for brief quotations for the purpose of review or comment and except for educational use (see below), no part of this book may be reproduced, stored in a retrieval system, or transmitted in any form or by any means—without the prior permission of the publisher, P&R Publishing Company, P.O. Box 817, Phillipsburg, New Jersey 08865-0817.

A teacher who purchases one copy of this book may photocopy the entire book up to five times for use in her own classroom without requesting permission. Charging for copies or otherwise using them for commercial use is strictly prohibited.

Page design by Tobias Design
Typesetting by Michelle Feaster

Printed in the United States of America

ISBN 0-87552-733-7

Contents

	To Parents and Teachers	7
	Meet the Author	9
1.	Sara	11
2.	A French Lesson	12
3.	Ermengarde	13
4.	Lottie	14
5.	Becky	16
	Test Your Knowledge	17
6.	The Diamond Mines	18
7.	The Diamond Mines Again	20
8.	In the Attic	22
9.	Melchisedec	25
10.	The Indian Gentleman	26
	Test Your Knowledge	28
11.	Ram Dass	29
12.	The Other Side of the Wall	31
13.	One of the Populace	32
14.	What Melchisedec Heard and Saw	34
	Test Your Knowledge	35
15.	Providence	36
16.	The Visitor	38
17.	"It Is the Child"	39
18.	"I Tried Not to Be"	40
19.	Anne	40
	Test Your Knowledge	42
	"Get Creative!" Activities	44
	Summary	54
	Answer Key	56

To Parents and Teachers . . .

Mark Twain once said, "A classic is something that everybody wants to have read and nobody wants to read." All too frequently this is true because the enduring power of the classics is lost in tedious classroom assignments. Books that could be savored become drudgery. To help you and your young reader avoid this pitfall, we encourage you to use this study guide wisely and creatively. Here are some issues to keep in mind:

- **Vocabulary Building.** Reluctant readers may benefit by studying the vocabulary words in advance of each reading assignment. This provides them with greater reading confidence and less distraction from the story. More gifted readers will know many of the vocabulary words or will deduce their meaning from the text itself.

- **Comprehension Questions.** Alternate between having students respond to questions orally and in writing. If a student can't answer a particular question, don't belabor it. Keep a list of the more challenging questions for further reflection and return to them later to see if understanding has blossomed. Don't expect a pat answer to every question. Many of the questions are intended to draw out personal opinions and theories, important skills for a maturing mind.

- **Story Continuity.** Do not interrupt an especially satisfying reading session by stopping at the end of each chapter to do the study guide exercises. Sometimes it is best to read several chapters at a time, saving the exercises for a natural pause in the flow of the plot.

- **Group Interaction.** The whole family or class can enjoy this classic story together. Whenever possible, adults

TO PARENTS AND TEACHERS . . .

and young readers should take turns reading pages or chapters aloud. Then spend a few minutes responding to the story and to one another's thoughts and feelings. Looking back on years of teaching, author and educator Aidan Chambers observes, "It was booktalk that pumped blood into our literary veins and gave us the energy, the impetus, for exploration beyond our familiar boundaries." Be sure to make time in your home and classroom for plenty of stimulating booktalk.

- **Creative Activities.** Intersperse vocabulary and comprehension building with the more expressive activities at the end of the study guide. These activities are designed to help students move from understanding the story to responding at a more personal level, a key element in internalizing and applying truth. Do not attempt to do all the activities; the number and variety is provided to allow teachers and students a wide range of choices.

- **Assignments.** Students are encouraged to keep a separate notebook for written answers to vocabulary and Think About It questions. This will maintain a permanent record of what they have learned from *A Little Princess*.

- **Answer Key.** There is an answer key and book summary for teacher use in the back of this book. The Knowledge Tests and Answer Key are to aid teachers in giving pop quizzes to test reading comprehension and recall. For definitions of vocabulary words, consult *Webster's New Collegiate Dictionary*.

As you begin this reading adventure, keep in mind these words of award-winning children's author Katherine Paterson: "The gift of creative reading, like all natural gifts, must be nourished or it will atrophy. And you nourish it [as] you read, think, talk, look, listen, hate, fear, love, weep—and bring all of your life like a sieve to what you read."

Meet the Author:
Frances Hodgson Burnett
(1849–1924)

Some people are born storytellers. Frances Hodgson Burnett certainly was. Even when she was a young child living in Manchester, England, her greatest pleasure was in making up stories and acting them out, using her dolls as characters.

Frances was the oldest of five children and daughter of a successful businessman who could provide his family with a nice home. But when she was four, her father died and the family suddenly became poor. Desperate to escape the factories and tenements of Manchester, they moved to a farming area near Knoxville, Tennessee. There Frances fell in love with nature, and wrote stories in a little attic room.

To help support her family, Frances started writing romance stories for women's fashion magazines. These stories sold well, but Frances grew tired of them and began to write about life as she had known it in England.

In 1873 Frances married her long-time friend Swan Burnett, a medical doctor. They soon had two boys—Lionel and Vivian—and went to live in Washington, D.C. There, on the third floor of a big brick house, Frances continued to write wonderful tales.

If you could talk to Vivian, here's what he would tell you about his mother's study: It was "a magic place, filled with all sorts of cozy feelings and comforts for her boys and her visitors. It had a fireplace, with a black fur rug before it, and on that a big armchair that held a sleepy boy easily on a mother's lap, even when he had become quite big and heavy."

Vivian later wrote a note to his mother saying, "You write so many books for grown-ups, that we don't have

MEET THE AUTHOR

any time at all with you now. Why don't you write some books that little boys would like to read? Then your staying upstairs wouldn't be so bad." His mother thought this was a wonderful idea. She started a children's story right away. Then every night she read to her sons from her new novel called *Little Lord Fauntleroy*. When this book was published in 1886, it became a best-seller. Right away Frances wrote another story for children, this time about a girl named Sara Crewe . . . *A Little Princess*.

Although Frances was not a regular churchgoer, she read the Bible regularly and cared deeply about spiritual issues. She said of herself, "Long ago, even when I was a little girl writing in the attic room in the Ark [her family's home] in Knoxville, I began to feel that I could not have it on my conscience to make people unhappy, or make their minds foul with anything I had imagined or put on paper. There is enough of that in our lives that we cannot get away from. What we all want is more of the other things—life, love, hope—and an assurance that they are true. With the best that was in me, I have tried to write more happiness into the world."

Chapter 1

SARA

PART 1: What's in a Word?
Match each word with its correct definition.

1. ___ thoroughfare
2. ___ vehicle
3. ___ bungalow
4. ___ resign
5. ___ destination
6. ___ seminary
7. ___ varnished
8. ___ intense
9. ___ intimate
10. ___ frock

a. strong
b. dress
c. accept; give in to something
d. school
e. means of transportation
f. goal; target
g. close; dear
h. cottage; small house
i. main road; highway
j. polished; painted

PART 2: Think About It . . .
1. Name three or four things about going to school in London that must have been hard for Sara. Which of them do you think was the hardest?

CHAPTER 1

2. Why did Sara like the world of grownups better than the world of children?

3. How were Miss Minchin and her house alike?

4. What made people think of Sara as "a little princess"?

Chapter 2

A FRENCH LESSON

PART 1: **What's in a Word?**

Look up the following words in the dictionary. Then use each one in a sentence.

1. whimsical
2. companion
3. engaged
4. imposing
5. annoyance
6. grave
7. exquisite
8. mortified
9. infuriated
10. grudge

PART 2: **Think About It . . .**

1. Why did Sara make believe that Emily was alive?

2. What made Miss Minchin think that Sara needed to learn French?

3. How does a child become "spoiled"? Do you think Sara was spoiled? Why or why not?

4. When Miss Minchin realized she had made a mistake, why did she become angry with Sara?

Chapter 3

ERMENGARDE

PART 1: **What's in a Word?**

Choose the best definition for each of the following words.

1. pigtail
   a. the end part of a pig   b. braided hair   c. rope
2. pathetic
   a. awful   b. excellent   c. ordinary
3. disdain
   a. admiration   b. scorn   c. appreciation
4. savage
   a. gentle   b. angry   c. frightened
5. titter
   a. laugh   b. whistle   c. stutter
6. disconsolate
   a. hopeless   b. generous   c. furious
7. intellectual
   a. smart   b. foolish   c. ignorant
8. profound
   a. shallow   b. deep   c. rescued
9. mournful
   a. satisfied   b. confused   c. sad
10. passage
    a. hallway   b. stairs   c. closet

CHAPTER 3

PART 2: **Think About It . . .**

1. If Ermengarde had been an excellent student, would Sara have become her friend right away? Why or why not?

2. Sara's father thought, if she had been a boy and born centuries earlier, she would have made a good knight. What caused him to think this?

3. How does Sara's attitude toward those in need compare to Jesus' attitude? (Hints: Matthew 9:36; Luke 7:13; 10:30–37; Hebrews 4:15–16; 5:2)

4. What is the difference between Sara's pretending and lying? Is it wrong for Sara to make believe?

Chapter 4

LOTTIE

PART 1: **What's in a Word?**

Use the following words to fill in the blanks in the sentences below them.

virtue	humiliation	amiable
temper	calculate	endeavor
apparent	flighty	convictions
eccentric	appalling	coax
alluded	indignation	grievance

1. My cousin is _____ and often changes her mind.

2. When the rain started, it soon became _____ that we couldn't go on our picnic.

3. It is a _____ to do what is right even when you don't feel like it.

4. When you have a _____ against someone, try to control your _____.

5. If you wear a bow tie to the baseball game, people will think you are _____.

6. Can you _____ the correct answer to the problem without any help?

7. After the fight, Tom's _____ was obvious to everyone.

8. The coach _____ to his disappointment with last year's team.

9. My sister has deep _____. No one can _____ her into doing something that she believes is wrong.

10. Please _____ to be more _____ when we have guests at the dinner table.

11. The amount of homework I have this weekend is _____.

12. Some people feel _____ when they aren't as well dressed or educated as their friends.

PART 2: Think About It . . .

1. What made Sara think that she might not be as good a person as she seemed? What does the Bible say about our goodness? See Isaiah 64:6.

2. How did Sara react to being Miss Minchin's star pupil?

3. Why did Sara want to help Lottie?

4. What was wrong with the way Miss Minchin and Miss Amelia treated Lottie?

5. How could you follow Sara's example by helping someone younger than you?

Chapter 5

BECKY

PART 1: **What's in a Word?**

Match each word with its correct definition.

1. ___ forlorn
2. ___ hearth
3. ___ replenish
4. ___ drudge
5. ___ grate
6. ___ grotto
7. ___ culprit
8. ___ scuttle
9. ___ stunted
10. ___ bower
11. ___ perplexity

a. make full; refill
b. retreat or cozy nook; clump of trees
c. trouble; hardship; tragedy
d. common folk; multitude
e. metal pail used to carry coal
f. one who does hard, boring work
g. small cave, often in or under water
h. confusion
i. metal bars that hold a fire
j. hallucination; dream
k. wrongdoer

CHAPTER 5

12. ___ calamity   l. miserable; sad
13. ___ delirium   m. small; undersized
14. ___ populace   n. fireplace

PART 2: **Think About It . . .**

1. What are two careers or occupations that might be just right for a grownup Sara Crewe?

2. Did Becky show that she was lazy by stopping her work to listen to the story and falling asleep in Sara's room? What makes you sure of your answer?

3. What did Sara mean when she said that her mother knew stories belong to everyone?

4. Why did Lavinia say that Sara was wicked? Was Lavinia right?

5. Sara told Lavinia she could find much more splendid stories in the book of Revelation. Where can you find this book? What does it say about heaven? (Hint: Revelation 4:1–6; 21–22)

6. What did Becky need just as much as food and rest? How did Sara give her what she needed?

7. Why did Sara wish she were a real princess?

Test your knowledge of Chapters 1–5 by answering these questions. (See back of guide for answers.)

1. Where did Sara live before she came to London?
2. What kind of work did Sara's father do?

CHAPTER 5

3. Where was Sara's mother?
4. Why did Ralph Crewe bring his daughter to London?
5. Who owned the boarding school Sara attended?
6. Who was Monsieur Dufarge?
7. How did Sara learn to speak French?
8. What first caused Miss Minchin to be angry with Sara?
9. Who was Sara's first friend at school?
10. Who was most jealous of Sara?
11. What did Sara pretend about Emily?
12. Besides Emily, who was Sara's little girl?
13. Why did the girls like to gather in Sara's room?
14. Where did Becky fall asleep?
15. Who was Mariettte?

Chapter 6

THE DIAMOND MINES

PART 1: **What's in a Word?**

Mark each sentence below with a ★ if the vocabulary word is used correctly and the sentence makes sense. Mark the sentence with an X if the word is used incorrectly.

___ 1. The people who wanted to get rid of their king started a **revolution**.

___ 2. The prisoner tried to escape from the **dungeon** but failed.

___ 3. Please **restrain** the milk into the carton.

CHAPTER 6

___ 4. Miss Minchin felt **rage** toward students who didn't do exactly what she wanted.

___ 5. We watched the balloon **ascend** into the sky.

___ 6. Ben was shy and fearful, so he volunteered for the **perilous** task.

___ 7. When a story is told in parts over a period of time, each story part is called an **installment**.

___ 8. I like **scuttling** on my bicycle.

___ 9. Lavinia was Sara's **benefactor** and did whatever she could to hurt her.

___ 10. Are you feeling **anxious** about tomorrow's test?

___ 11. Our teacher planned **festivities** for the pupils who misbehaved.

___ 12. I went to the butcher shop for a **garland**.

___ 13. When Sara was **mystified**, she understood exactly what was happening.

___ 14. My baby brother wears warm, fuzzy pajamas made of **flannel**.

PART 2: **Think About It . . .**

1. When Captain Crewe wrote to Sara about the Indian diamond mine, she thought of a story from the Arabian Nights. You probably know at least one character from these famous stories—Aladdin. Another important character is Sindbad, a merchant and sailor. To read the story Sara had in mind—*The Valley of Diamonds*—visit your library and look for *Tales from the Arabian Nights* by C. Lang, edited by Andrew Lang (NTC/Contemporary Publishing, 1998) and *Sindbad* retold and illustrated by Ludmilla Zeman (Tundra Books, 1999). Also, you may want to read the DK Classics "Eyewitness" version of *Aladdin* by Rosalind

Kerven and Nilesh Mistry (DK Publishing, 1998). It is filled with illustrations that will help you understand the exotic world Sara told about in her stories.

2. Like Lavinia, you may find it hard to imagine a mine filled with diamonds. To most of us, it sounds like a fairytale. But diamond mines are real. To learn more about diamonds and how they are mined, read *The Story of Diamonds* by Jean Milne (Shoestring Press, 2000). If your local library doesn't have this book, ask a librarian to get a copy for you through interlibrary loan or to suggest a similar book.
3. Is it possible to be both a beggar and a princess? Why or why not?
4. In what way is every child of God a princess or prince? (Hint: 1 Peter 2:9; 1 John 3:1; Revelation 1:5–6)
5. How did thinking of herself as a princess affect the way Sara treated Lottie? Becky? Lavinia?
6. The author tells us that God had made Sara a giver—a person with open hands and an open heart. Reread the first half of page 56. Then think about the people you know who are givers. Make a list of them. How could you let them know you are thankful for them?

Chapter 7

THE DIAMOND MINES AGAIN

PART 1: **What's in a Word?**

Match each word with its correct definition.

1. ___ jest
2. ___ scandalize

a. forlorn; sad
b. coin worth one quarter of a penny

CHAPTER 7

3. __ murmur   c. beggar; street person
4. __ garret   d. pass off onto someone else
5. __ irate   e. kindness; generosity
6. __ farthing   f. whisper; complaint
7. __ foist   g. awaken; excite
8. __ shrewd   h. memory
9. __ sumptuous   i. smart; sly
10. __ preposterous   j. furious; extremely angry
11. __ pauper   k. horrify; shock
12. __ desolate   l. bully; push around
13. __ grotesque   m. odd-looking; ugly
14. __ recollection   n. expensive; magnificent
15. __ charity   o. attic; upstairs storage area
16. __ domineer   p. joke
17. __ surge   q. rush; flow
18. __ rouse   r. ridiculous; absurd

PART 2: **Think About It . . .**

1. Why was Becky glad to stay at Sara's party in spite of the terrible way Miss Minchin treated her?

2. The Bible says, "Pride goes before destruction, a haughty spirit before a fall. Better to be lowly in spirit and among the oppressed than to share plunder with the proud" (Proverbs 16:18–19). How did Miss Minchin set herself up for a fall?

3. What kept Miss Minchin from putting Sara out on the street?

4. How did Sara keep Miss Minchin from feeling as powerful as she wanted to feel? Was Sara's behavior wrong?

5. Why did Becky's kindness cause Sara to cry?

6. When Sara's father died, her suffering was too great for words. For a long time she felt numb, as though her own life had ended. Perhaps you have lost someone you love and understand how Sara felt. If you haven't, you may sometimes fear that you will. Look at pages 36–37 and 41 to find out what comforted Sara and gave her hope. Remember what you read in Revelation 4:1–6; 21–22. Also, you and your family may want to read one or two of the following books:

- *Someday . . . Heaven* by Larry Libby (Zondervan, 2001)
- *What Happens When We Die?* by Carolyn Nystrom (Moody, 1992)
- *Let's Talk about Heaven* by Debby Anderson (Chariot Victor, 1998)
- *Someone I Love Died (Please Help Me, God)* by Christine Harder Tangvald (Chariot, 1988)

Chapter 8

IN THE ATTIC

PART 1: **What's in a Word?**

Choose an **antonym** *(word that means the opposite) for each of the following words.*

1. intense
   a. strong   b. weak   c. warm

CHAPTER 8

2. numb
   a. feeling   b. unfeeling   c. paralyzed
3. handmaid
   a. servant   b. helper   c. housewife
4. loath
   a. unwilling   b. eager   c. afraid
5. besiege
   a. attack   b. protect   c. confuse
6. encounter
   a. meet   b. introduce   c. avoid
7. wretched
   a. miserable   b. mean   c. happy
8. injustice
   a. wrong   b. unfairness   c. fairness
9. listless
   a. alert   b. dull   c. lazy
10. loiter
    a. delay   b. hurry   c. dawdle
11. affectionate
    a. hateful   b. loving   c. thoughtful
12. stoutly
    a. weakly   b. stubbornly   c. bravely
13. enrapture
    a. delight   b. disgust   c. please
14. adversity
    a. hardship   b. trouble   c. prosperity

CHAPTER 8

PART 2: **Think About It . . .**

1. Several times the author mentions Sara's proud heart. Is it different than Miss Minchin's pride? If so, how?

2. Sara did her best to be a good worker. How was she treated in return?

3. Why did Miss Minchin want to keep Sara away from the other girls?

4. Sara pretended she was a soldier in a war. In fact, she was—in the battle between good and evil. What does the Bible say about how to be a good soldier in this war? (See: Ephesians 6:10–18; 2 Timothy 2:1–4)

5. Because of Ermengarde's visit, Sara saw a possible reason that God had allowed her to suffer. What was this reason?

6. Do you think Sara was right when she said there might be good in things even when we can't see it? Can you give an example of this? (Be sure to read Romans 8:28 and Hebrews 12:11.)

7. When Sara decided to turn her trials into adventure stories, she thought about stories she had read. One of these famous stories is *The Count of Monte Cristo* by Alexander Dumas. If you want to read this exciting story look for the edition by Robin Waterfield (Puffin, 1996). Sara decided her attic room would be a cell in the Bastille. You can learn about this famous French prison by looking in the encyclopedia, or check out Charles Dickens's *A Tale of Two Cities* in an edition for young readers (abridged by Linda Jennings for Puffin, 1996).

CHAPTER 8

8. In her talk with Ermengarde, Sara said, "Adversity tests people, and mine has tested you and proved how nice you are." How does adversity test people? (For some clues, read James 1:2–4 and 1 Peter 1:6–7. You can read about the test Ermengarde passed in Proverbs 17:17.)

Chapter 9

MELCHISEDEC

PART 1: **What's in a Word?**

Write five sentences correctly using all of the following words.

1. attentive
2. accountable
3. aghast
4. existence
5. hob
6. apologetic
7. occupant
8. possession
9. countenance
10. distinct

PART 2: **Think About It . . .**

1. How did Sara make her attic room seem like a wonderful place to Lottie?

CHAPTER 9

2. Sir John Lubbock once said, "Everyone must have felt that a cheerful friend is like a sunny day, which sheds its brightness on all around; and most of us can, if we choose, make of this world either a palace or a prison." In what ways was Sara spreading sunshine and turning her prison into a palace?

3. Why was it important for Sara and Becky to talk with each other through the wall?

4. When Sara said everything and everyone is a story, what did she mean?

Chapter 10

THE INDIAN GENTLEMAN

PART 1: **What's in a Word?**

Choose the best answer from the choices below.

1. If you make a **pilgrimage**, are you going:

    a. to bed   b. to dinner   c. on a journey

2. When your mother makes an **inspection**, is she:

    a. checking out a book   b. checking your room
    c. writing a check

3. Would a **rare** coin be:

    a. common   b. unusual   c. worthless

4. If you are **clad**, are you:

    a. dressed   b. undressed   c. poorly dressed

5. When something **invariably** happens, is it:

    a. frequent   b. without fail   c. unexpected

CHAPTER 10

6. If you have a **sixpence**, do you have:

   a. money   b. jewelry   c. a good luck charm

7. Would a **dismayed** person:

   a. laugh   b. gasp   c. give you a compliment

8. When your father is **profoundly** interested in what he is reading, does he:

   a. fall asleep   b. turn on the television   c. ignore interruptions

9. Where would an animal go to **nestle**?

   a. to its mother   b. up a tree   c. in the garage

10. If you **console** your friend, will you:

    a. tease   b. comfort   c. warn

11. When someone behaves in a **vulgar** way, is the person:

    a. nasty   b. nice   c. graceful

12. Is a **vacant** house:

    a. empty   b. full   c. for vacations

13. What might be a sign that someone feels **remorse**?

    a. rage   b. giggles   c. improved behavior

14. If your home is **elaborate**, is it:

    a. simple   b. comfortable   c. fancy

15. When a piece of furniture is finely **wrought**, is it:

    a. old-fashioned   b. foreign   c. well-made and decorated

16. Would you be most likely to find a **heathen**:

    a. in a church   b. at a pagan ceremony   c. reading the Bible

17. What do people do with their **idols**?

    a. worship them   b. ignore them   c. insult them

CHAPTER 10

18. If your grandmother shops for **figurines**, is she looking for:

   a. idols   b. statues   c. cookies

19. When someone offers you a **tract**, should you:

   a. read it   b. run down it   c. follow it

20. You would be more likely to see a **haggard** face on a:

   a. model   b. prisoner   c. milk carton

PART 2: Think About It . . .

1. What was the hardest part of Sara's new life?

2. Why did Sara make a necklace out of the sixpence she received rather than spending it?

3. Sara believed it was a sign of great strength to keep from losing her temper. Do you agree? What does the Bible say? (Hints: Psalm 39:1; Proverbs 16:32; 17:27; 25:28; 29:11; James 1:19–20)

4. When Sara's pain became almost more than she could stand, something happened to bring her new hope. What was it? Has this ever happened to you during a difficult time in your life?

5. Becky believed their new Indian neighbor was a heathen, or idol worshiper. Did she have any proof? She thought someone should send him a tract about Christianity. Was this a good idea without getting to know him?

Test your knowledge of Chapters 6–10.

1. Captain Crewe's friend asked him to become a partner in a new business. What was it?

CHAPTER 10

2. Where did Becky live?
3. What funny name did Sara's father call her?
4. What did Becky give Sara for her birthday?
5. Who came to visit Miss Minchin during Sara's birthday party?
6. What happened to Captain Crewe?
7. Why did Miss Minchin threaten to expel Sara from her school?
8. Where was Sara's new room?
9. What made Ermengarde miserable?
10. Who was Melchisedec?
11. What did Sara like about her new room?
12. Which family in Sara's neighborhood did she like best?
13. What did a little boy give Sara?
14. Why did Sara knock Emily onto the floor?
15. Where was Sara's new neighbor from?

Chapter 11

RAM DASS

PART 1: **What's in a Word?**

Match each word with its correct definition.

1. ___ splendor   a. light up
2. ___ vaulted   b. lengthen; stretch out
3. ___ enclose   c. mischievous
4. ___ molten   d. hot liquid

CHAPTER 11

5. ___ alter
6. ___ illuminate
7. ___ impish
8. ___ restore
9. ___ accustomed
10. ___ precaution
11. ___ prolong
12. ___ agile
13. ___ interval
14. ___ triumph
15. ___ execute
16. ___ malicious
17. ___ insolent; impudent
18. ___ civility
19. ___ personage

e. put to death
f. in the habit of; used to something
g. boldly rude
h. courtesy; politeness
i. change
j. grandeur; brilliance
k. quick and nimble
l. make something new; return
m. arched; dome shaped
n. hateful
o. victory; success
p. surround
q. space of time; pause
r. care taken to prevent harm
s. individual; human being

PART 2: **Think About It . . .**

1. What did Sara and Ram Dass have in common?

2. How did Ram Dass encourage Sara to keep acting like a princess?

3. What made Miss Minchin treat Sara in such a different way than Ram Dass treated her? What made her blind to Sara's splendid character?

4. Why did Sara apologize for laughing, but refuse to apologize for thinking?

Chapter 12

THE OTHER SIDE OF THE WALL

PART 1: **What's in a Word?**

Use the following words to fill in the blanks in the sentences below them.

misfortune	cease	schemes
imperiled	torment	swindlers
disgrace	absorb	sane
sympathy	catastrophe	conscious
harass	reproached	vague

1. Grandmother feels great _____ for us when we have problems. She doesn't like to see us suffer _____.

2. Watch out for _____! They have many _____ to take your money.

3. Don't _____ or _____ your brother unless you want to be _____ for your bad behavior.

4. The flood was a major _____ and it _____ many lives and homes.

5. When I hear a new idea, sometimes it takes me a while to _____ it.

6. Our neighbor was _____ about when he will fix the hole in the fence.

7. When someone is sick with a high fever, he may not be sensible or _____.

CHAPTER 12

8. Are you _____ that you only have five minutes left to finish the test?

9. It is a _____ to turn your back on a friend when she is in trouble.

10. When will this cold, rainy weather _____ so that we can go outside and play?

PART 2: **Think About It . . .**

1. Sara enjoyed watching the large, happy family that lived near her. She "adopted" them, claiming them as her friends even though she didn't know their names. Have you ever adopted someone as a friend without ever speaking to them? If you have, describe your experience.

2. When Sara sent kind thoughts to her neighbor from India, what was she really doing? Why did she feel better afterward?

3. Who was Mr. Carmichael and why was he looking for Sara Crewe?

4. Have you ever run away from a problem and felt bad about it later as Mr. Carrisford did? Where can we find the courage to face our problems? (Hint: Psalm 56:3; 138:3; Isaiah 41:10.)

Chapter 13

ONE OF THE POPULACE

PART 1: **What's in a Word?**

1. Name a place where you might **tramp**.

2. What does the sky look like on a **dismal** day?

CHAPTER 13

3. Where would you find a **subterranean** treasure?
4. Name a group of people who are **downtrodden**.
5. What do you most hate to be **deprived** of?
6. In what room of the house are you most likely to find **currants**?
7. What are three things families keep in the **cellar**?
8. Where do we see **pavement**?
9. Name a place where people **jostle** one another.
10. When you are **ravenous**, what do you do?
11. Where is your school **situated**?
12. Name three things that **hover**.

PART 2: **Think About It . . .**

1. Sara said, "What you have to do with your mind when your body is miserable is to make it think of something else." Have you ever tried this? Did it work? Describe your experience. When one of God's people named Asaph went through a terrible time, he found a way to get his mind off his troubles. Read Psalm 77:11–12 to find out what he did, and try this the next time you feel miserable.

2. Why did Sara ask the woman in the bakery if she had lost a fourpence?

3. What caused Sara to give away almost all the buns she bought?

4. Why did the woman in the bakery treat the child on her doorstep differently after Sara left?

Chapter 14

WHAT MELCHISEDEC HEARD AND SAW

PART 1: **What's in a Word?**

Look up the following words in the dictionary. Then use each one in a sentence.

1. mystified
2. invade
3. dexterity
4. outcast
5. mounted
6. thrust
7. emerge

PART 2: **Think About It . . .**

1. What were Ram Dass and the Indian gentleman's secretary doing in Sara's room?

2. Was it wrong for Ram Dass to spy on Sara? Why or why not?

3. In India, Ram Dass's home, people are divided into groups called *castes*. If a person is born into a high caste, he or she has many privileges. But those born into the lowest caste have almost no opportunity to succeed in life. Which caste did Ram Dass think Sara should belong to?

CHAPTER 14

4. Did Ram Dass and Mr. Carrisford believe in the same god? How can you tell?

Test your knowledge of Chapters 11–14.

1. Who did Sara see poking his head up from the skylight next door?
2. What animal lived in the house next door to Miss Minchin's school?
3. How did Sara's neighbor get into her room?
4. What did Sara do in the evenings after she finished her chores?
5. Was the owner of the house next door Indian or English?
6. Who was Mr. Tom Carrisford?
7. What was wrong with Mr. Carrisford?
8. Who was Mr. Carmichael and what was his job?
9. What did Sara buy with the money she found?
10. Who received a very special gift from Sara?
11. Where did Mr. Carmichael go to look for Captain Crewe's daughter?
12. Who made a point of learning everything he/she could about Sara, even spying on her in her room?
13. What terrified Melchisedec?
14. What did Mr. Carrisford plan to do for Sara?
15. Who was Sara's fellow prisoner in what she called "the Bastille"?

CHAPTER 15

Chapter 15

PROVIDENCE

PART 1: **What's in a Word?**

If you correctly match all the words below with their meanings, you deserve a special reward!

1. ___ vicious
2. ___ lopsided
3. ___ flush
4. ___ rueful
5. ___ gory
6. ___ squire
7. ___ vassal
8. ___ clarion
9. ___ minstrel
10. ___ hospitality
11. ___ falsehood
12. ___ clench
13. ___ passion
14. ___ barrier
15. ___ hampers
16. ___ transform
17. ___ insignificant
18. ___ contort
19. ___ flagon
20. ___ alabaster

a. slave; serf
b. bloody; frightening
c. warm welcome; generosity
d. unimportant
e. trumpet
f. cup; mug
g. lie
h. twist in a strange way
i. God's help
j. intense feeling or emotion
k. full of regret
l. baskets
m. musician
n. stringed musical instrument
o. knight's servant
p. candles
q. woodwind instrument
r. grip; hold tightly together
s. blush; turn red or pink
t. quick and graceful in moving about; nimble

CHAPTER 15

21. ___ bonbon
22. ___ tapers
23. ___ damsel
24. ___ viol
25. ___ bassoon
26. ___ lithe
27. ___ providence
28. ___ flyleaf

u. lady
v. wall; hurdle
w. change completely
x. mean; hateful
y. off balance; uneven
z. white marble
aa. candy
bb. blank front page in a book

PART 2: **Think About It . . .**

1. Why didn't Ermengarde realize that Sara was starving?

2. What kept Sara from asking Ermengarde to bring her food?

3. What bothered Sara about making Ermengarde's father think that his daughter had read the books he sent?

4. What did Sara believe was more important than being smart or clever?

5. Look in the encyclopedia to find out about Maximilien Robespierre, the villain Sara described to Ermengarde. (Online, go to: http://www.comptons.com/encyclopedia/ARTICLES/0150/01556480_A.html/)

6. Have you ever been falsely accused of something the way Becky was accused of stealing? How did it feel? What does God say about false accusations? (Read Proverbs 6:16–19; 19:5; 25:18.)

7. What did Sara say always happens before things get to the very worst? (See pages 152 and 163.) Is this true? (Psalm 46:1; 1 Corinthians 10:13; Hebrews 4:15–16.)

CHAPTER 15

8. Why was it hard for Becky to "see" the things Sara imagined? What can you learn from Sara about exercising or strengthening your imagination?

9. Why did Sara cry when she saw what was written on the flyleaf of her new book?

Chapter 16

THE VISITOR

PART 1: **What's in a Word?**

Choose the correct answer from the following choices.

1. If you **falter**, do you fail or only hesitate?
2. When you **luxuriate** in an experience, do you enjoy it or dislike it?
3. Is a **deceitful** person trustworthy or dishonest?
4. If your teacher looks **askance** at you, is he suspicious or pleased?
5. Would you feel **elated** if you won a race or if you lost it?
6. When we treat people with **severity**, are we harsh or kind?
7. If you experience **deprivation**, do you have plenty or not enough?
8. Where would you put something that you want to **conceal**: in everyone's sight or in a hiding place?
9. Do most Americans enjoy many or few **conveniences** in their homes?
10. Does an obedient or a disobedient child show **defiance**?
11. If you become **agitated**, are you upset or peaceful?
12. Would a **crotchety** person speak gently or crossly?

CHAPTER 16

PART 2: **Think About It . . .**

1. Why did Miss Minchin wish Sara would be impudent or break down from the punishment she received?

2. Is it wrong to want to break a child's spirit as Miss Minchin wanted to break Sara's? (Hint: Proverbs 18:14; Matthew 18:1–6, 10; Ephesians 6:4)

3. What is the difference in defiance and the way Sara acted?

4. What made Mr. Carrisford think Sara would be allowed to wear the clothes he sent her? Was he right?

5. What did Sara believe generous people wanted more than thanks?

6. Who could you make happy by writing a thank-you note?

7. Why did Sara sign her note as "The Little Girl in the Attic" instead of Sara Crewe? What would have happened immediately if she had signed her real name?

Chapter 17

"IT IS THE CHILD"

PART 1: **What's in a Word?**

See if you can write a single sentence that makes sense using all four of the following words.

1. astride
2. misery
3. reluctant
4. contracted

PART 2: **Think About It . . .**

1. How was Providence at work as Mr. Carmichael, Mr. Carrisford, and Ram Dass talked?

2. Why did Mr. Carrisford say he could not bear to question Sara?

Chapter 18

"I TRIED NOT TO BE"

Chapter 19

ANNE

PART 1: **What's in a Word?**

Choose a **synonym** *(word that means the same) for each of the following words.*

1. reproachful
    a. fault-finding   b. approving   c. disrespectful
2. presumption
    a. rudeness   b. politeness   c. medicine
3. involuntarily
    a. intentionally   b. instinctively   c. rudely
4. proprietress
    a. owner   b. teacher   c. dressmaker
5. scrutiny
    a. examination   b. fairness   c. hatred

6. intrude

   a. dodge   b. bother   c. shove

7. significance

   a. importance   b. meaning   c. both a and b

8. flinch

   a. flutter   b. cringe   c. scowl

9. heir

   a. inheritor   b. pretender   c. leader

10. undertake

    a. refuse   b. adopt   c. accept

11. grovel

    a. pebble   b. snarl   c. beg

12. penetrate

    a. fill   b. pierce   c. smash

13. accomplish

    a. do   b. neglect   c. fail

14. in readiness

    a. forgotten   b. prepared   c. ignored

15. proposal

    a. denial   b. plan   c. engagement

PART 2: Think About It . . .

1. Why was Sara confused at first about how she should feel toward Mr. Carrisford?

2. What did Sara's experience have in common with Psalm 23:5?

3. Where did the title of chapter 18 come from?

4. How did Miss Minchin's relationship with her sister Amelia change?

5. What made Becky happy and sad about Sara's new fortune?

6. Did Sara's character change after Mr. Carrisford rescued her? How can you tell?

7. What rule did Sara follow in her daily life? (Hint: Luke 6:31)

8. How did Sara's kindness inspire others to be kind?

9. Who is your example of how to treat people?

10. If you were a character in this story, who would you be? Why?

Test your knowledge of Chapters 15–19.

1. What two exciting gifts did Ermengarde bring to Sara?

2. What color were Sara's eyes and hair?

3. Why did Miss Minchin tell Becky she deserved to be sent to prison?

4. Who told Miss Minchin that Ermengarde was sneaking up to Sara's room?

5. What happened on the same night Miss Minchin caught the girls having their party in the attic?

6. As Sara grew older, what was Miss Minchin's plan for her?

7. Who sent Sara two packages? What was inside the packages?

CHAPTERS 18-19

8. What did Sara leave for her new friend on the table in her room?
9. Why did Sara keep the runaway monkey in her room all night?
10. How did Mr. Carrisford discover Sara's identity?
11. Who came looking for Sara when she disappeared from school?
12. Who told Miss Minchin that she was a hardhearted woman?
13. What wonderful surprise came to Becky after Sara left the school?
14. What kind of pet did Mr. Carrisford give Sara?
15. How did Sara want to spend part of her fortune?
16. What surprise waited for Sara at the bakery?

"Get Creative!" Activities

Use an encyclopedia and other library books to learn about the two places that were important to Sara—India and London. Then divide a poster board in half. Decorate one half of the poster to illustrate Sara's life in India. Decorate the other half of the poster to illustrate her life in London. You can use pictures cut from old magazines (not from the library, please!), stickers, and drawings you make. Include buildings, foods, clothing, animals, busy streets, musical instruments . . . anything that a person might see in India or London. Your poster will help you better understand Sara's two different worlds.

The English language borrows words from many other languages, including Hindi. See if you can find the meanings of the following Hindi words in *A Little Princess*: *ayah*, *bungalow*, *lascar*, and *sahib*. Once place to look is on the Internet at http://www.an-daman.org/book/app-i/texti.htm. Look through the list of words you find there and see if you can identify other words like *bungalow* that have become a part of our English language.

"GET CREATIVE!" ACTIVITIES

Plan a tea party around the theme of *A Little Princess*. It can be very simple like the party Sara gave with the goodies from Ermengarde's gift box. Or your tea party can be like the one provided by Mr. Carrisford and Ram Dass. Either way, use your imagination to make it special and fun. You'll find recipes for your party in *The Secret Garden Cookbook* by Amy Cotler and Prudence See (HarperCollins, 1999).

Sara said, "If nature has made you a giver, your hands are born open and so is your heart. And though there may be times when your hands are empty, your heart is always full, and you can give things out of that." Make a list of fifty specific ways you can give to others when you don't have money. Then try them!

Remember how Sara befriended the lonely little girl named Lottie? Is there a young child who needs your attention? Maybe you could read the story of Sara Crewe to her or him. Look at your library for the picture book version of *A Little Princess* edited and illustrated by Barbara McClintock (HarperCollins, 2000). If you can't find this book to read to your little brother, sister, or friend, make your own illustrations and tell the story in your own words.

Sara discovered, "Whatever comes . . . cannot alter one thing . . . I can be a princess inside." Write three paragraphs in which you describe someone you know who is a princess or prince on the inside.

"GET CREATIVE!" ACTIVITIES

Like Princess Sara, Frances Hodgson Burnett was a born storyteller. Even when she was a young child living in Manchester, England, her greatest pleasure was in making up stories and acting them out, using her dolls as characters.

You can act out Sara Crewe's story using *A Little Princess Paper Dolls* by Judith Sutton (HarperCollins, 1999; available from a local bookstore or online at www.amazon.com). Before you cut out the paper dolls, take time to glue them onto lightweight cardstock (available at stationery or craft stores) using an acid-free glue stick (available at craft stores). When the glue has thoroughly dried, then cut out the figures.

When Sara found a silver fourpence coin near the bakery, she went inside and bought buns. Here is a recipe for the wonderful buns she shared with the beggar girl on the bakery doorstep. Ask your parents or grandparents if you can have them for a special treat. Be sure to help in the kitchen before and after the baking. Also, think about whether there is someone you could cheer up by sharing this tasty treat.

Currant Buns

1/4 cup lukewarm water
1/4 cup granulated sugar
1 pkg. active dry yeast (1 tbsp.)
3 1/2 cups all-purpose flour
1/2 tsp. salt
1 cup milk
1/4 cup butter, melted
1 egg
1/2 cup currants or golden raisins

"GET CREATIVE!" ACTIVITIES

Icing:

2 tbsp. powdered sugar
1 tbsp. water

1. In a small bowl or measuring cup, combine warm water and 1 tbsp. of the sugar. Sprinkle the yeast over the top of the water and sugar mixture. Then let these ingredients stand for 10 minutes or until frothy.
2. Warm the milk and the butter until the butter melts. Then whisk a beaten egg into this mixture.
3. In a large bowl, blend together the remaining sugar, flour, and salt.
4. Make a well in the dry ingredients. Pour the yeast mixture into the well. Then add the milk, butter, and egg mixture. Use a wooden spoon to stir until a soft dough forms. Last add the currants, making sure that they are distributed throughout the dough.
5. Cover the bowl and let the dough rise for about 2 hours, or until it is double in size.
6. Place the dough on a lightly floured board. Knead for 8 minutes or until the dough is smooth and flexible.
7. Shape the dough into a 12-inch long roll. Cut the roll into 12 equal pieces.
8. Use your hands to shape the pieces of dough into smooth, dome-shaped balls.
9. Place the buns on greased baking sheets, leaving about 2 inches between them. Cover the baking sheets loosely with waxed paper and leave them in a warm place for 30 minutes (or until the dough has doubled in volume).
10. Bake the buns in a 400-degree oven for 15 minutes or until golden brown.
11. While the buns bake, mix together powdered sugar and water to form icing. This icing can be brushed onto the buns just before they are done or while they are still warm from the oven.

"GET CREATIVE!" ACTIVITIES

During childhood, many famous people loved *A Little Princess*, including writer Phyllis McGinley. Now that she is grown, she still thinks it is a wonderful book because "it has everything a good tale ought to have—a hero and a villain (or, rather, a villainess) and a princess in disguise (for Sara is really a princess because she is so noble) and a mysterious benefactor and magic and many adventures and a happy ending. It is much better than Cinderella, which it resembles, because it could have happened."

How many books have you read that contain all the ingredients Phyllis McGinley named? Make a list of them. Next, choose your top three favorites and write a short description of each one. Then, ask several of your friends to do the same thing. See if any of your choices match theirs. After you hear one another's book descriptions, there may be some new books you want to read.

Have you ever thought about how stories change over the years? For example, before Frances Hodgson Burnett wrote *A Little Princess*, she wrote a shorter story called "Sara Crewe." Another famous writer named Mary Mapes Dodge (author of *Hans Brinker*), published "Sara Crewe" in a children's magazine called *St. Nicholas*. The story was so popular that it became a book the next year with the title *Sara Crewe or What Happened at Miss Minchin's*. In 1902 and 1903 the story became a play and was seen by audiences in London and New York. Frances Hodgson Burnett added several characters to the story in order to make the play. This made her want to tell more about Sara's life. So, two years later, she gave readers *A Little Princess: Being the Whole Story of Sara Crewe Now Told for the First Time*. More recently there have been several movie versions of *A Little Princess*.

"GET CREATIVE!" ACTIVITIES

Check with your video store and library for the following movies about Sara Crewe:

- *The Little Princess*, starring Shirley Temple (1939)
- *A Little Princess, Parts 1–3*, staring Amelia Shankley and Maureen Lipman (WonderWorks Public Media, 1990; 3 videos)
- *A Little Princess*, staring Eleanor Bron and Liam Cunningham (1995)

How are the movie versions different from the book? What is good about the movies? What do you dislike about them?

Imagine that you are director of a new movie version of *A Little Princess*. What actors would you choose for the following roles?

1. Sara Crewe
2. Ralph Crewe
3. Miss Minchin
4. Becky
5. Ermengarde
6. Lottie
7. Lavinia
8. Ram Dass
9. Mr. Carrisford
10. Mr. Carmichael

Have you ever dreamed of becoming a writer? Do you think that you are too young? You definitely aren't too young! Frances Hodgson Burnett began to write stories when she was just seven years old. If you truly want to write, choose a place to work. Then go there every day and write, even if it's just for five or ten minutes. In or near your writing space, keep pens and pencils, paper, crayons or pens if you like to illustrate your stories, and a dictionary. If there is no space you can set aside for your writing, then fill a basket or plastic tote box with what you'll need. That way you can grab it quickly and head for a quiet place.

Here is some advice for young writers from Frances Hodgson Burnett:

1. Ask yourself why you want to write. Do you just want to be the author of a book? That isn't enough. You need to be curious about life and caring toward people. You need to spend time trying to understand why people do things and why they don't. If this sounds like you, give it a try.

2. Remember that no matter how unimportant something may seem, it means joy or grief to the person who experiences it. "*Everything* is a human story."

3. Read and think about the best writing the world has known and loved.

4. Do your best to find the right words for what you want to say.

5. Tell stories of hope and encouragement. There is enough in our lives to make us unhappy or to make our minds feel dirty. As Frances said, "What we all want is more of the other things—life, love, hope—and an assurance that they are true."

(Source: *Happily Ever After: A Portrait of Frances Hodgson Burnett* by Constance Buel Burnett; Vanguard Press, 1965.)

Some people are good at putting stories on paper. Other people, like Sara Crewe, are great at telling stories

"GET CREATIVE!" ACTIVITIES

to an audience. A good place to begin telling stories is at home, and there's a book that will help you and your family have more fun with stories. It's called *The Family Storytelling Handbook: How to Use Stories, Anecdotes, Rhymes, Handkerchiefs, Paper and Other Objects to Enrich Your Family Traditions* (by Anne Pellowski; Macmillan, 1987). Ask for it at the library.

There are dozens of storytelling groups and events where people gather to listen to stories and to learn how to tell them well. To find out more, you and your parents can go to your favorite Internet search engine (like www.google.com) and type in the word "storytelling."

Sara's doll Emily held a very special place in her heart. Have you ever had a toy that was like a friend? Write a story about it.

Miss Minchin was guilty of many wicked things. Although none of them broke the laws of London, they broke God's law. List the laws of God that Miss Minchin broke (See Exodus 20:1–17 and James 2:8). What could Sara and Mr. Carrisford do to protect other girls from her cruelty? Help them make a plan that will work.

Did you ever wonder why Sara and Becky stayed at Miss Minchin's school when she treated them so badly? Why didn't they run away? If you think about how hard life was for Anne, the little girl Sara saw on the bakery doorstep, you'll know why they stayed at the school. To better understand what life was like for homeless children in London, read:

"GET CREATIVE!" ACTIVITIES

- *Oliver Twist* by Charles Dickens, abridged by Ian Andrew and illustrated by Naia Bray-Moffat (DK Publishing, 1999)
- *The Christmas Doll* by Elvira Woodruff (Scholastic, 2000)

When you finish these stories, look for some ways you and your family can help homeless people in your city.

Sometimes authors write a series of books so that we know what happens to characters as they grow up. Since Frances Hodgson Burnett didn't do that, we have to use our imagination. What do you think might have happened to Sara and the people in her life in five years? ten years? twenty years? Write a paragraph or two about the future of each main character.

Two Extra Special Challenges

You are a detective. See if you can find out why Ralph Crewe was a soldier in India. Why did his country send him there? What was his job? What was it like for British families who lived in India around the end of the 19th century? Why did many of them send their children to boarding schools in England? How did the Indian people feel about having British soldiers in their country? To answer these questions, you'll need help. Ask your parents, teacher, and librarian where to find answers. Check in the encyclopedia, books about India, and on the Internet. Here are three clues to help you begin your search: British Raj, Colonial India, and British Empire.

"GET CREATIVE!" ACTIVITIES

Read the Old Testament story of Joseph (Genesis 37, 39–45, if possible in a children's Bible). Compare Joseph's life to the story of Sara Crewe. How many things can you find that they had in common? If you and your family or class have read *A Little Princess* together, you may want to work together on this activity.

A Little Princess
Summary for Parents and Teachers

A Little Princess is the story of Sara Crewe, a wealthy young student at a London boarding school. When tragedy suddenly strikes, Sara finds herself at the mercy of the cruel schoolmistress, Miss Minchin. Overwhelmed by terrible trials, Sara must find the strength to survive. She discovers hope in a wonderful secret—a secret that magically transforms even the lowliest of beggars into true royalty. Sara's discovery is that, "Whatever comes . . . cannot alter one thing. If I am a princess in rags and tatters, I can be a princess inside." Her spirit remains unbroken in spite of grief, suffering, and persecution.

This story is a gold mine of values, encouraging readers toward perseverance, bravery, generosity, and imagination. It is an enduring classic because of the captivating and entertaining way Burnett portrays triumph over adversity. One reviewer describes *A Little Princess* as a "child-sized depiction of a battle between good and evil—with the good and evil both so unusually interesting. Sara isn't a goody-two-shoes, but a thoughtful and creative person who is genuinely interested in the meaning of right and wrong; the fact that doing the right thing doesn't always come easy to her makes her accomplishments all the more admirable."

Most of the people around Sara, particularly Miss Minchin, judge a person's worth by external appear-

SUMMARY FOR PARENTS AND TEACHERS

ances. Her story provides readers with a vivid illustration of the biblical principle that true worth is a matter of the heart. In the midst of a "what's-in-it-for-me" society, Sara is a giver. Burnett writes of her, "If nature has made you a giver, your hands are born open and so is your heart. And though there may be times when your hands are empty, your heart is always full, and you can give things out of that." In our materialistic, consumer-oriented culture, the importance of this message for young readers cannot be overstated.

Phyllis McGinley, author of *Plain Princess, Sixpence in Her Shoe,* and *The Most Wonderful Doll in the World,* writes, "This book is just about the most interesting, funny, sad, exciting, wonderful story anybody ever told." Although much has changed since *A Little Princess* was written, McGinley observes, "Courage doesn't change. Mysterious good deeds do not change. There are still wicked people and good people and brave unhappy little princesses left in the world."

Ranelda Hunsicker, author of this study guide, met little princess Sara Crewe when she was nine years old and they quickly became best friends. Sara and Ranelda had a lot in common—fathers whose work took them away, love of books and beautiful things, financial hardship, and mistreatment by unkind people. Sara taught her about courage, kindness, determination, and hope. From Sara, Ranelda learned to believe that dreams really do come true, that God and His human helpers somehow always find ways to deliver delightful surprises into the dark places of our lives.

Kathryn Lindskoog has edited this timeless classic for today's reader, faithfully preserving every detail of the original stories while featuring an accelerated pace, updated terminology, and full-page illustrations to add to your reading pleasure.

KNOWLEDGE TESTS ANSWER KEY

Most of the questions in this study guide are purposely designed to encourage reflection, discussion, and written response. For these kinds of questions, it is impossible (and counterproductive) to provide pat answers. However, to aid you in creating pop quizzes, here is a key to chapter summary items in the guide.

Test for Chapters 1–5
1. India
2. British military officer
3. Dead, or in heaven
4. To go to school
5. Miss Minchin
6. The French teacher
7. From her mother, who was French
8. Her mistake about whether Sara knew how to speak French
9. Ermengarde
10. Lavinia
11. That Emily was a real girl instead of a doll
12. Lottie
13. To hear Sara tell stories
14. In Sara's room
15. Sara's French maid

Test for Chapters 6–10
1. A diamond mine
2. In a room in the attic
3. Little Missus
4. A handmade pincushion
5. Captain Crewe's attorney (lawyer)
6. He died of brain fever.
7. Because she was a penniless orphan
8. Next to Becky's room in the attic
9. She thought Sara didn't want to be her friend.
10. A rat who lived in Sara's attic room
11. The skylight
12. The Large Family (also called the Montmorency family)
13. A sixpence

14. Because she wanted someone to talk with so badly and Emily couldn't talk
15. India

Test for Chapters 11–14
1. Ram Dass
2. A monkey
3. By crossing from one house to the other on the rooftops
4. She studied by herself.
5. An Englishman from India
6. Sara's neighbor
7. He was sick from brain fever and worry.
8. Mr. Carmichael was the man Sara thought of as Mr. Large or Mr. Montmorency. He was Mr. Carrisford's attorney.
9. Currant buns
10. The starving beggar girl on the bakery doorstep
11. Russia
12. Ram Dass
13. The coming of Ram Dass and Mr. Carrisford's secretary into Sara's room

14. Make her room beautiful and provide her with food and clothes
15. Becky

Test for Chapters 15–19
1. Books and food
2. Green eyes and black hair
3. For stealing food from the kitchen
4. Lavinia
5. Ram Dass brought all sorts of wonderful things to Sara's room.
6. For Sara to teach in the school without pay
7. Mr. Carrisford sent Sara two packages filled with nice clothes.
8. A thank-you note
9. It was cold and snowy outside.
10. From things Sara said when she returned the monkey to him
11. Miss Minchin
12. Miss Minchin's sister Amelia
13. Sara asked Becky to move next door and be her maid.
14. A dog
15. Food for the poor
16. The beggar girl Anne had found a good home.